ANIMAL HOMES

L. Robinson

BRIAN WILDSMITH

Oxford University Press

OXFORD NEW YORK TORONTO MELBOURNE

Oxford University Press, Walton Street, Oxford OX2 6DP

Oxford is a trade mark of Oxford University Press

© Brian Wildsmith 1980

First published 1980
Reprinted 1983, 1990, 1991
First published in paperback 1991

ISBN 0 19 279732 8 (hardback)
ISBN 0 19 272173 9 (paperback)

Printed in Hong Kong

The snail lives in its shell.
When there is danger, it
hides itself completely
inside the shell. As the snail
grows, the shell gets bigger.
When winter comes, the
snail goes to sleep, safe
inside its shell.

The armadillo likes the dark. It digs into the ground with its strong claws and makes a burrow for itself. At night it comes out of its burrow and eats insects. The armadillo is covered with an armour of bony scales from head to tail.

Yaks live in the high mountains of central Asia and Tibet. They are very good at climbing rocks. Yaks have a very thick coat to protect them from the cold winds in the mountains.

The walrus lives in the Arctic Ocean,
either on the shore or on ice-floes.
It eats shellfish, which it digs off the
rocks and the sea-bed with its long tusks.

Beavers live in rivers and lakes.
They build lodges from tree trunks
and branches. In the autumn
they plaster the lodge with mud
to keep the cold out.

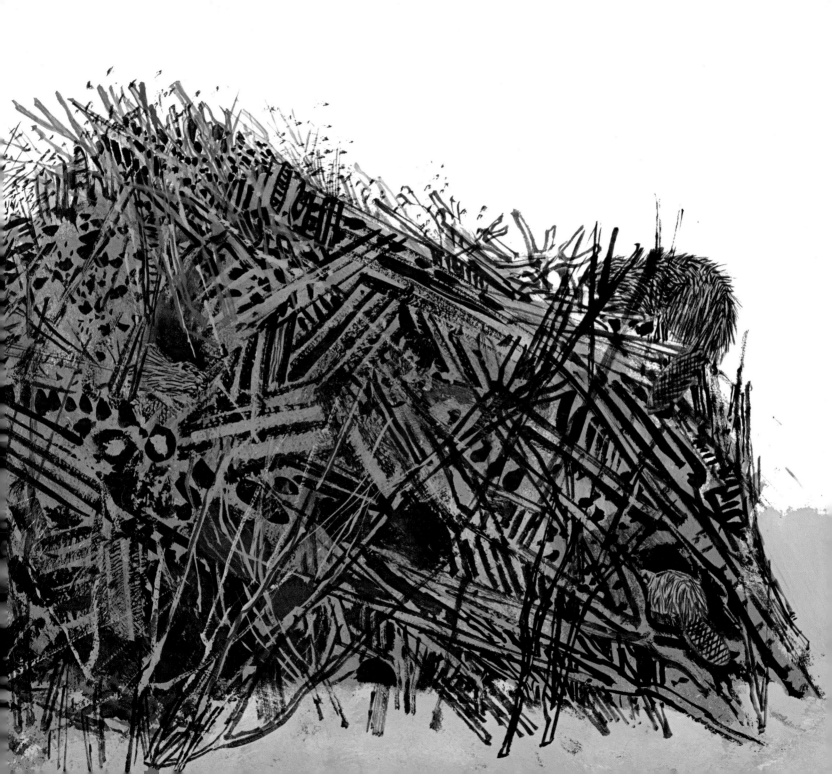

An eagle's nest is called an eyrie.
The eagle makes it from sticks and
builds it high on a mountain or on
a tree-top. The eagle always uses the same eyrie,
and every year makes it bigger and bigger.

The chameleon lives in
trees and bushes, and can
change its colour to
hide itself. It catches insects with
its long, sticky tongue.

Koalas only live in eucalyptus trees
in Australia. They spend all day
sitting in a tree, eating leaves.
The mother carries her baby
in her pouch for three months,
and then carries it on her back.

The wolf makes its lair in a rocky cave,
a hole in the ground, or in the hollow
of a fallen tree-trunk. Wolves hunt by
day and by night. In winter they always
hunt in packs.

Kingfishers live in burrows which they dig with their beaks in the banks of a stream. They catch fish for food. They skim across the water and then dive down to spear the fish with their beaks.

You cannot tame a wild cat.
It lives mostly in the mountains,
and makes its home in a hollow tree.
When the female has kittens,
she puts them in a nest
in a hole away from the male.
Otherwise he might kill them.

Brown bears live in rocky caves in the mountains. Their home is called a den. Bears can grow very big, but they are not fierce and will not harm anyone unless they are angered.